Nicholas G. Stangarone

NWP and the Energy Enlightenment Manifesto.

Gold Mind Press

NWP and the Energy Enlightenment Manifesto

Copyright © 2021 by Nicholas G. Stangarone

All Rights Reserved. No part of this book may be reproduced or transmitted in any form or by any means, electronic or mechanical, including photocopying, recording, or by any information storage and retrieval system without permission in writing from the publisher.

ISBN-13: 9798759187301

Printed in the United States of America

First Printing 2022

This book is dedicated to Mr. Nello Ori. In gratitude of his splendid advocation of my beleaguered circumstance all those years ago.

Contents

Introduction *page 11*

CHAPTER ONE
The Energy Enlightenment Envisioned. *page 15*

CHAPTER TWO
Putting the Pieces of Power in Place. *page 25*

CHAPTER THREE
Virtuous Power Is Universal. *page 39*

CHAPTER FOUR
Shock the Monkey. *page 47*

CHAPTER FIVE
A Launch For The Ages. *page 53*

Appendix 1.......... *page 59*

Appendix 2.......... *page 73*

Appendix 3.......... *page 85*

Appendix 4.......... *page 92*

Introduction

This book begins not with the first chapter, but rather with the last chapter of the first book titled, "Ring of Power," that laid initial claim to the marvelous effects of "leveraged momentum." For in that final chapter the concept of *"**Energy Enlightenment**"* was touched upon, but not fully addressed.

This book is an attempt to accomplish this to a point where a solid base of reasoning and logic can be established. And then from this real and substantiated beginning proceed further still toward the greater goals that are to be addressed in later chapters.

I would venture to speculate that when our prehistoric ancestors were first inspired to control fire and utilize the energy released by burning wood or other organic substance to heat their shelters and cook their rancid meats and coarse vegetables into delectable and nourishing meals that something close to a religious epiphany may have also been experienced.

This profound capability to possess the awesome divinity of the pagan gods, I believe began the process that leads us to this day. For way back when, even before the Stone Age, humans and neanderthals alike, spent the bulk of their daily efforts desperately avoiding becoming the next tasty meal for some or other ferociously hungry carnivore. While likewise, these primitive bipeds were hunting smaller, much more docile reptile or mammal for their lunch. So when fire gave them the power to overcome their biological fragilities, some form of ecclesiastical/symbiotic relationship developed. And it is that power which has brought us to the point where man's collective efforts at energy production and consumption is actually destroying the only planet that is even remotely livable.

For example, In 2020, over hundreds of ancient giant sequoia trees were destroyed by fire. These awe-inspiring trees, some a millennia old had certainly experienced innumerable times of exposure to incredibly treacherous weather extremes. Yet, through countless centuries, these noble living conifers had survived and even thrived.

Well that was, until now. So basically what humans have accomplished during their brief existence on this planet has been to take a unbelievably vast myriad of natural life-sustaining ecologies and reduced them to throughly artificial landscapes. And all borne of man's diabolical aspirations to control everything.

So, whether we like it or not the time has come for drastic changes to the status quo of take, take, take it all. The time has come to get real and personal with how we produce and utilize our energy, specifically our electricity.

However, what that actually entails has yet to be entirely determined. And is some small but yet tangible way this book hopes to shed some light on what awaits out there in the distant darkness.

So, the first chapter will present an overview of the New World Power Philosophy, including the Energy Enlightenment. Chapter two will cover several technologies that will be utilized with this potential power source. The third chapter will layout in brief detail some of the promised benefits that will help humanity reached its solemn destiny.

Chapter four will cover "Shock the Monkey" Light Extravaganza and awards show to celebrate "NWP and the Energy Enlightenment."

And the fifth and final chapter will cover some of the ways to put the "Energy Enlightenment" philosophy into effect.

Chapter One

The Energy Enlightenment Envisioned.

Imagine, if you will, what the earliest humans envisioned when they stared into the dazzling brilliance of a roaring camp fire and felt it's radiant heat. It had to be truly awe-inspiring, this power they had captured to bring light and warmth to their precarious lives.

Now, picture in your mind that almost instantaneously combustive explosion that pressures a steel piston, precisely fitted in a cylindrical barrel to spin a crankshaft that effectively propels a ponderous but luxurious SUV toward the occupant's intended destination.

One flame gives life-giving sustenance to prehistoric man while the other allows modern man to travel effortlessly across time and space.

One flame is crude and haphazard while the other is

precisely measured and ignited by spark for maximum thermodynamic output. Yet both of these flames produce the same thermodynamic effects. The flame is always the same, now and even at the time of the "Big Bang."

And once the human species had harnessed the power of the flame, they had unknowingly transcendent the animal domain. For while animals in the wild will always suffer silently the extremes of nature, man could now alter his fabricated environment to accommodate his daily need to shelter from the storm.

Fire cooked their raw, rancid meats and coarse vegetables into edible, nourishing food. Besides heating their homes in the cold of winter, fire could help bake durable pottery and render lethal weaponry. And some probably realized, metaphysically speaking, they had also captured a piece of the burning sun.

Significantly, in Greek mythology, Prometheus is a Titan god best known for defying the gods by stealing fire from them and giving it to humanity in the form of technology, knowledge, and more generally, civilization.

But poor Prometheus is mentioned only because this fire myth is so universal. But fire or through fire, man has always seen or spoken to their God. There's the burning of bush that's depicted in the bible. The Zoroastrian's eternal fire and the Vestal Virgin's vigil of their famous scared flame. Even the Latin word for their priests was flamen. So it is not beyond reason to postulate that combustion of fuel was a profound religious phenomena to prehistoric man.

And by which logic, so was the phenomena we now call electricity. Especially in the appearance of lightening and thunder. These natural meteorological occurrences were also deified in prehistoric times and pre-Christian pagan era. Most notably Zeus and his Roman counterpart Jupiter and also the Nordic deity, Thor.

For our highly intelligent pre-historic ancestors, no doubt had recognized or come to believe in the religious and magical significance of fire and electricity. And this itself would eventually evolve to include some moral or ethical context.

Yet the modern human in all their technological splendor and modernity, have fully failed to grasp the sacred nature of energy. Modern man has effectively separated any vestige of sacredness or holiness from energy and turned it into a supply/demand market commodity.

But NWP does not look on the timeless wisdom of the ancient philosophers as anarchistic to modernity and the cult of technology. In fact, NWP considers ancient philosophy and religious practices an extension of the ageless understanding of the undeniable differences between the virtues and vices of man.

And taking this another step further, NWP considers the lack of effort or input demonstrated by the global religious hierarchy over the means by which energy is produced and consumed as an abdication of the God-given right to offer theological insight and wisdom over the whole spectrum of human interaction.

And without attempting to be overly critical, one could

argue that it is because of this fact and in addition to other corrupting influences that have lead to our current climate calamities. All religious authority have hindsight of historical fact to reclaim their sacred right to oversee how energy is used or wasted. At least follow Pope Francis' lead and begin to offer some form of moral and ethical guidance that is so sorely missed in these modern times.

Why was comprehending the moral distinctions between virtue and vice so crucial to the development of a just and honorable character? Is not the actions of either virtue or vice but the existential presence of our eternal divinity of soul? Especially in comparison to the animal kingdom, where most actions are instinctual? That is, they act on extreme hunger and seasonal hormonal overloads.

But before proceeding further with what's has been mostly conjecture, let me just state what is a fundamental precept of the Energy Enlightenment: "All able-bodied adult humans are bequeath by Earth to generate sufficient energy necessary to sustain a virtuous life."

Here, for clarification's sake, allow me to reiterate that it is only by man's greed that they, or more specifically the global oil and energy conglomerates have purposely deprived us of this highly enriching energy.

But equally as important is the understanding that if Earthlings are made capable of generating suitable quantities of electrical power, then they would not have "necessity" as an excuse for allowing the hideous practice of knowingly and willfully depriving future inhabitants of Earth their rightful claim to the planet's depleted resources.

I mean think about it, what sort of monstrous global society have we become when we accept that it is somehow ethically and morally just to force future Earthlings to escape their planet for a more suitable planet because we had made it inhabitable for them due to our insatiable greed for money and power.

Don't you think it would be more worthwhile to make Earth a more environmentally life-nurturing planet. You know, exactly how planet Earth was before "man the monster" came along and totally **"FOULED"** everything up.

But unfortunately this once avoidable calamity will not only harm the perpetrators, but also all of Earth's other unique and priceless life forms. Earthy manifestations of life which had a certain right to live it's natural course of being as any living entity. But man, once chosen by their almighty GOD had been given the right to destroy all by the flame. To put all that lives to torch. Because that is our right by fire.

So here we have our moment of truth; man's moral conundrum: Everything man touches is tainted by moral consequence.

And it is this inconvenient truth which now bring us to comprehending the great epiphany of the Energy Enlightenment.

However, before we examine the tenets of the Energy Enlightenment more concisely, I would like to briefly postulate how I came to observe the moral wave of continuity that permeates every aspect of energy production; from how the lucrative one-sided deals are struck to how the crude oil is wastefully and violently suck out of Earth and then finally burned off in an instant with nothing remaining but environmental pollution and additional thermal pollution.

While other personally conducted experiments and also related academic studies have most certainly played a role into the insights and attitudes I currently hold to be true. Yet, if one where inclined to fix a particular event as precipitating what followed, I would probably conclude that it all began with the exercise invention I had devised called the **GYROCISOR**.

And the reason for this is definitely because the **GYROCISOR** physical fitness equipment specifically utilizes the dynamic kinetic forces of sweeping rotational motion in an elliptical trajectory during the execution of the **GYROCISOR** "Double-Eagle" Superstroke. And since 2004 when I built the first **GYROCISOR** "Power Shuttle," I've been exercising with one on a regular basis over the course of nearly two decades.

But aside from becoming the primary workout of my weekly fitness regime, this exercise had also rather curiously, afforded the opportunity to observe by grasp (touch) the highly dynamic kinetic forces resulting from the angular momentum of a weighted "Power Shuttle." This "hands on" approach to appreciating the many nuances of the orbiting weight allowed me to capture within my calculating mind the constant motion as a static physical quantity. And it was at that instant that the whole universe opened itself to this brief but profound epiphany of force.

Not dissimilar from how Galileo had observed and studied the motion of rolling balls on an inclined plane. Understanding the mathematical implications of the inclined plane's angle of attack to the speed of the accelerating motion of free falling objects had presented a readily discernable and therefore measurable quantity. But whereas Galileo's observations were purely mathematical. The experiments I conducted were predominantly "hands on" experiments and imaginatively intuitive.

Yet it was not until this next invention that the kinetic dynamism of angular momentum could first be leveraged into potentially producing satisfactory amounts of electrical power. But of even far greater consequence, it was the "Ring of Power" that has provided proof that angular momentum could be effectively leverage, thereby laying the groundwork to support NWP's vision quest and the Energy Enlightenment's idealistic tenets.

So now, let's see exactly what these tenets of the Energy Enlightenment.

1.) How we acquire energy (Super-Fungible Commodity)* has moral consequence that can resonate throughout the soul collective like ripples across a pond's surface. But in this circumstance it's across today's global economy.

For example, if treachery and deceit are used during any point in the energy production process, especially at the onset, these morally caustic and highly corruptible actions permeate across the energy supply chain to the final destination as work output.

2.) If we damage or alter the environment or it's natural living inhabitants to produce energy, this will have long-lasting consequences that cannot be determined at that present time and perhaps for generations to come.

3.) If we consume or destroy the energy resources of earth's future inhabitants we should be judged as having had committed crimes against future humanity..

4.) God does not deprive men from producing their own life-sustaining energy, it is only because of man's insatiable greed that we find themselves dispossessed.

This is because Earth has bequeath each of it's inhabitants the source to eternal energy.

But just as man must have specially developed equipment to fly through earth's atmosphere. Man must have specially developed equipment to produce substantial energy. And this where ROP technology can offer a means to fulfill the NWP vision quest.

*The definition of "fungible" goods are items that are interchangeable because they are identical to each other for practical purposes. Commodities, common shares, options, and dollar bills are examples of fungible goods. Energy is being termed "super-fungible" here because it also just as readily transfers moral or ethical prerogatives. Whether just and virtuous or otherwise.

So now that we've begun to establish some of founding principles of the Energy Enlightenment, let's further clarify what New World Power is all about.

In many ways, NWP is to act as a beacon of illuminating intelligence cast across an endless ocean of dark and unknown waters toward realizing a truly just and equitable balance to man's energy resources and consumption.

It will be through ongoing efforts like those proposed by NWP as inspired by the EE that profound change can be ultimately achieved. But before proceeding any further with the methods and tactics to be utilized, let's proceed to the chapter two and consider some of the important elements that will play a part in the grand movement.

Chapter Two

Putting the Pieces of Power in Place.

New World Power and especially the Energy Enlightenment may seem abstract, or at least derived of abstract conceptions. However concrete strategies and tactics to achieve these tendered claims and objectives have been prepared. So let's review some technological capabilities that will be utilized once the operations have commenced.

All of the information presented below has been researched from numerous sources. Yet, it is important to note that the predominant share has been provided by **wikipedia**. And all I will further say about this is that **wikipedia** is really an incalculably valuable service and which by my personal estimation is certainly worthy of a Nobel Peace Prize.

And with that bit of recognition, let's now proceed with eight features of the ROP (Ring Of Power or ROP is presented here as a generic term not unlike the term, Refrigerator.) that are important to the realization of the NWP vision quest and the Energy Enlightenment.

#1) Alternators (EMF/Induction)

#2) Robotics

#3) A.I., Artificial Intelligence:

#4) VR, Virtual Reality

#5) Video Gaming

#6) Cryptocurrency

#7) Marijuana Cultivation:

#8) 5G

#1) Alternators (EMF/Induction)

The alternator must truly be considered one of the greatest inventions that has ever been devised throughout the history of mankind. I believe it is akin to when man first discovered how to harness the power of fire. To the ancient mind, today's portable generators would have been tantamount to stuffing the gods Zeus or Jupiter into a small box. And then, simply by turning the crank on this wonder box, a mere mortal could force the gods to immediately reply with endless bolts of lightning. How absolutely shocking, indeed!

Obviously, the modern mind has been educated well enough to satisfactorily comprehend devices such as batteries which store electricity by chemical energy or generators and alternators which produce electricity by converting rotational mechanical energy into an electrical current.

Of course, with generators and alternators this force is the result of electromagnetic induction or the production of an electromotive force across an electrical conductor in a changing magnetic field.

Just for some historical context, it's been nearly two hundred years since Michael Faraday's discovery of electromagnetic induction in 1831. And since that time, so very long ago, the physical phenomena of electromagnetic induction is used in an astoundingly array of applications, including electrical components such as inductors and transformers, and devices such as electric motors and generators.

Obviously, modern gas-powered automobiles are completely reliant on compact alternators mounted to its engines for charging the battery and to power its electrical system. Other types of alternators and generators that are used for diesel train engines, wind power, hydro-applications, marine alternators for boating and much more that are well beyond the scope of this book.

Finally, one other crucial component of NWP's power grid will require adequate power storage, though not just chemical like the current (lead-acid) batteries but also other different types of batteries will be used. These could include pneumatic (compressed gases,) and flywheel energy storage.

#2) Robotics

When I was born almost smack dab in the middle of the twentieth century, robots were mostly in science fiction movies. But today in the early twenty-twenties that's no longer the case. They have become nearly ubiquitous, though not as imagined. No, they may not be made in our image, but they are now hard at work somewhere. Because the majority are predominantly utilized as "assembly robots."

But there are as well other applications such as military robots, industrial robots, Cobots, construction robots and agricultural robots. And there are also medical robots such as the da Vinci Surgical System.

Then there's robot combat for sport, cleanup of contaminated areas, such as toxic waste or nuclear facilities, Nanorobots, Swarm robotics and of course, autonomous drones. Robots are also being used in kitchen automation like the Zume Pizza, Cafe X, Makr Shakr, Frobot and even the Boris, a dishwasher loading robot.

Whether these robots will ever be used for fulfilling the NWP vison quest is not being pursued at this time. However they're worth mentioning because all the above mentioned robots require electrical power to operate.

#3) AI., Artificial Intelligence

Artificial intelligence (AI) is intelligence that is demonstrated by machines versus the natural intelligence displayed by humans and animals, which involves consciousness and emotionality. AI research includes reasoning, knowledge representation, planning, learning, natural language processing, perception and the ability to move and manipulate objects. Many tools are used in AI, including versions of search and mathematical optimization, artificial neural networks, and methods based on statistics, probability and economics. The AI field additionally draws upon computer science, information engineering, mathematics, psychology, linguistics, and philosophy.

So far during this century, AI techniques have experienced a resurgence following concurrent advances in computer power, large amounts of data, and theoretical understanding; and AI techniques have become an essential part of the technology industry, helping to solve many challenging problems in computer science, software engineering and operations research.

High-profile examples of AI include autonomous vehicles (such as drones and self-driving cars), major publishers now use artificial intelligence (AI) technology to post stories more effectively and generate higher volumes of traffic.

However, one should never underestimate the potential hazard that AI could become a danger to democracy if not humanity if it progresses unabated. Also AI, unlike previous technological revolutions, will create a risk of mass unemployment.

NWP feels strongly that AI must only augment the human intellect, not replace it as the final authority.

AI and Robots lack the ability to truly feel emotional and physical pain. And more importantly these machines can never be programed to do so because it revolves around life and the eternal soul. This is because AI and Robots cannot viscerally experience death as it relates to eternity and that any life once lost can never be recovered. Robots and computers can always be replaced with a new and improved version.

And since AI applications cannot, by definition, successfully simulate genuine human empathy, therefore AI research will consequently devalue human life. What's this? No new and improved clone of you? What a bleeding shame.

Now so called "ethical machines" will require policies for regulating artificial intelligence and robotics. However research in this area including machine ethics, artificial moral agents, friendly AI are years from realization and adaptation.

But still the concept of artificial moral agents (AMA) has become a part of the research landscape of artificial intelligence as guided by its two crucial questions which W. Wallach identifies as:

A.) Does Humanity Want Computers Making Moral Decisions?

B.) Can Robots Truly Be Moral?

Again, artificial intelligence has been briefly discussed here because it too requires constant electrical power to calculate and predict outcomes.

#4) VR, Virtual Reality & the Metaverse

A virtual reality headset is a head-mounted device that provides virtual reality for the wearer.

They comprise a stereoscopic head-mounted display, stereo sound, head motion tracking sensors and some also have eye tracking sensors and gaming controllers.

Virtual reality (VR) headsets are widely used with video games but they are also used in other applications, including simulators and trainers. Some examples would be a soldier in combat training using a VR headset or as a means to train medical students for surgery of any of those wounded in combat. Or more specifically, VR "augmented reality" headsets are also instrumental in the advancement of image-guided surgery.

#5) Video Gaming

A video game is an electronic game that involves interaction with a user interface or input device - such as a joystick, controller, keyboard, or motion sensing device - to generate visual feedback for a player. They are defined based on their platform, which include arcade games, console games, PC games, smartphones, tablet computers, virtual reality headsets and remote cloud gaming. They are also classified into a wide range of genres based on their type of gameplay and purpose.

The first video games were simple extensions of electronic games using video-like output from large room-size computers in the 1950s and 1960s, while the first "consumer" video games appeared in 1971. Since the 2010s, the commercial importance of the video game industry has been increasing along with the emerging Asian markets.

There are also serious Video games where the entertainment factor is secondary or eliminated to suit the intended purposes for that game. Educational games are a form of serious games, but other types of serious games include fitness games that incorporate significant physical exercise to help keep the player fit. Newsgames aimed at conveying a specific advocacy message.

#6) Cryptocurrency

The first true cryptocurrency, Bitcoin was invented in 2008. Shortly after it was released to the public as open-source software. Bitcoin is a decentralized digital currency, without a central bank or single administrator, that can be sent from user to user on the peer-to-peer bitcoin network without the need for intermediaries. Transactions are verified by network nodes through cryptography and recorded in a public distributed ledger called a blockchain. Bitcoins are created as a reward for a process known as mining.

Mining is a record-keeping service done through the use of computer servers and consumes a great amount of electrical power in the process. Miners keep the blockchain consistent, complete, and unalterable by repeatedly grouping newly broadcast transactions into a block. Each block contains a SHA-256 cryptographic hash of the previous block, thus linking it to the previous block and giving the blockchain its name. To be accepted by the rest of the network, a new block must contain a proof-of-work (PoW).

However since Bitcoin's inception, thousands of similar cryptocurrencies have come to full fruition. And if the cryptocurrency market continues to expand and evolve at its current pace, it will rival the global financial markets in scope and scale before the decade is out.

And as is now becoming common knowledge, the mining of any cryptocurrency consumes vast amounts of electrical power.

#7) Marijuana Cultivation

Today, legal cannabis can be grown outdoors, either on natural soil or in pots of pre-made or commercial soil.

To generate optimum quantities of THC-containing resin, the plant needs a fertile soil and long hours of daylight. In most places of the subtropics, cannabis is germinated from late spring to early summer and harvested from late summer to early autumn.

Outdoor cultivation is common in both rural and urban areas. Outdoor cultivators tend to grow indica-based strains because of its heavy yields, quick maturing time, and short stature.

Some growers prefer sativa because of its clear-headed (cerebral) high, better response to sunlight, and lower odor emissions.

In the Northern Hemisphere, growers typically plant seeds in mid-April, late May, or early June to provide plants a full four to nine months of growth. Harvest is usually between mid-September and early October.

And most importantly, the proceeds from any cultivation are intended to supplement the energy generating operations.

#8.) 5G Telecommunications

In telecommunications, 5G is the fifth generation technology standard for broadband cellular networks, which cellular phone companies began deploying worldwide in 2019, 5G networks are predicted to have more than 1.7 billion subscribers worldwide by 2025.

Due to the increased bandwidth, it is expected the networks will not exclusively serve cellphones like existing cellular networks, but also be used as general internet service providers for laptops and desktop computers, competing with existing ISPs such as cable internet, and also will make possible new applications in "Internet of Things" and machine to machine areas.

The main advantage of the new networks is that they will have greater bandwidth, giving higher download speeds, eventually up to 10 gigabits per second (Gbit/s). 5G speeds will range from ~50 Mbit/s to over a gigabit/s. The fastest 5G is known as mmWave. As of July 3, 2019, mmWave had a top speed of 1.8 Gbit/s on AT&T's 5G network.

And Now, How It All Comes Together.

#1) Alternators (EMF/Induction)

Basically an alternator or generator is an energy converter. The key to energy conversion is that the realm of the electron can be infinitesimally small or cosmically immense. And with that in mind one can say that the alternator's capabilities have yet to be fully realized and this is where NWP wants to lead the way. That is, to where EMF & I, Leverage and Momentum are united and a whole new world of opportunity opens.

#2) Robotics & A.I.

Robots and A.I. are combined here because both rely on electricity to perform their chosen tasks.

Since these technological wonders are purposely designed to perform beyond a human's range of skill and intelligence.

Therefore NWP proposes that it would behoove humans to concentrate on performing their superior skills of generating electricity to feed their power-hungry techno counterparts. And thereby completing the circle of "power and performance."

Then when the robots and A.I. are given the electrical power to perform expertly in their tasks, humans are mentally and physically empowered to cleaning up their "one and only" planet Earth.

#3) VR, Video Gaming and 5G

VR and Video gaming work hand in glove to provide a mind enhancing medium for a full-spectrum enjoyment of the digital content. And 5G will be the conduit to access all the mind-expanding digital content there is out there.

#4) Cryptocurrency

Whatever "excess power" is generated will be utilized for mining cryptocurrency. The operative word here is "excessive." Otherwise the power will be allocated for more immediate needs and uses.

#5) Marijuana Cultivation

Because NWP only advocates the production and use of clean, renewable energy. Any available land will be utilized for the cultivation of hemp and marijuana. The Ring of Power is one prime example on how ample open space can be made available for the cultivation of either hemp or marijuana.

38

Chapter Three

Virtuous Power Is Universal.

Chapter two reviewed the "instruments of change" and also offered a few brief points as to how these could be coordinated into a cohesive movement. But the ultimate goal is not to bring maximum ROI to a group of wealthy shareholders. NWP is a throughly legitimate attempt to accord a new balance that is both sustainable and equitable between all divisions of society. NWP is about bringing truth and justice to how energy is procured, distributed and ultimately consumed.

The Energy Enlightenment is the inevitable offshoot of what is an unalterable determination to bring a more ethically virtuous existence for all the human inhabitants that call Earth home.

So it's only natural that the initial effort should be made to assist those greatest in need. Or let's preface that with this would be the case with a morally mature and virtuous society. Yet with humans, greed always has an corrosive manner of opting for the unethical approach in return for the allure of cold, hard cash. And loads of it! And why would that be the case, you might purposely inquire with some glint of sincerity. My reply would be a one word answer: **INSATIABLE**.

Try to understand that the problem stems from the fact that greed, in and of itself, can never be satisfied. Greed is all about wanting more without regard to reason or moral implication. That is the whole gist of what greed means or represents. The purposeful negligence or unwillingness to consider the inevitable ramifications of never knowing when to stop wanting more and more and more, money of course.

Good and honest people must not be bullied by greed junkies and their tyranny of capitalistic lies and half-truths. They possess a God-given right to shield themselves from the soul-corroding effects of unfettered greed. After all, only a fool or miscreant would consider it a good idea to leave a fox guarding the hen house. So why do we allow this in our society?

I truly believe that one of the most astonishing things that I've personally witnessed was how easily a large proportion of the "Baby Boomer" generation got sucked into the pleasure vortex of "never-ending" want and desire. And all because they were born, or at least Americans such as my peers, into a sense of inherited entitlement that was buoyed by the euphoria of conquering the axis powers in 1945 and becoming the preeminent global economic superpower.

Suddenly finding themselves, cock of walk, young Americans were groomed for the intoxicating global aspirations of unlimited wealth and power. And while Americans did succeed in transforming the world into it's own consumer image. In the process, however, it has ravaged this planet's life-nurturing environments far beyond historically biblical proportions.

Humans now finds themselves at the precipice of life-altering global climate transformation. And to the degree that the outcome could have detrimental consequence is certainly debatable. But it is equally certain that great and profound change is at hand.

This is why it's now **TIME** for a change.

NWP believes that there are a minimum of four pillars to supporting a well-founded society.

These are adequate and safe housing, nutritious food, general fitness including overall medical and finally extended education and training. This in not intended to preclude other activities or pursuits of luxury, but these four cornerstones must be maintained at all costs for all its inhabitants. Regardless of race, religion, age or sexual orientation.

A.) Adequate and safe living quarters:

One of the ways that NWP proposes to significantly mitigate homelessness is through the Teepee Restoration

Program or TRP. This housing program is in recognition of the ingenuity shown by the indigenous people of America with their iconic teepee. Though this form of living space may not be the most practical in certain situations, NWP believes strongly in bringing the beautiful teepee back to the American landscape.

The reason why this is the first consideration is because the current situation in affordable housing is an ongoing, slow-motion societal disaster that can only get worse unless radical changes to the real estate market as a whole are instituted.

My days of living in California are a living testament to the ongoing debacle growing worse all across the American landscape.

Obviously there's always the tendency to blame the victim in this shameful tragedy. Yet those who have profited most and therefore are most to blame by default are made out as conquering hero's. And though they have won no "real" battle, they are allowed to take all the "spoils of war." And this is doubly true, because we are in a global economic battle and real estate is now proxy in effect to the global money wars.

This can no longer be tolerated by a morally righteous society.

Also, as part of the "Shock the Monkey" celebration event, which will be covered later, a teepee promotion to raise awareness for the "Wounded Knee Massacre" victims will be a very important aspect of the program.

B.) Affordable Nutritious Food:

Nutritious food is the basis for good health and mental fortitude. That is why NWP believes it to be of the highest crucial consequence to provide the freshest fruits, vegetables and nuts. Like housing, this is a must have requirement for any society to claim its democracy rests on sound foundation.

C.) Affordable Medical Care Includes Daily Health & Fitness Requirements.

This means ready access to medical specialists, team physicians and physical trainers. The most important criteria here is in prevention. That means taking the time and making the effort to stay and keep healthy through healthy living habits. This will require rewarding those activities and nutritional meals which are proven to prolong health and fitness.

D.) Provide Extended Education and Training

Though it is not NWP's objective to become an accredited college or university. However, it is worth nothing that NWP fully realizes that the EE relies on scientific research and other sources of knowledge or wisdom to increase understanding amongst the global population. So it is important to offer remedial education to supplement university curriculum.

Topics of importance would be history/philosophy, ethics and even languages. Not to be excluded would be mathematics and the sciences. And also Art history and techniques.

In addition, NWP will offer training in courses related directly to NWP's stated goals and tactical machinery and equipment.

These educational goals taken in aggregate do represent one particular reason why VR, Video Gaming and 5G are important to the overall NWP vision quest.

Chapter Four

Shock the Monkey.

Since its very inception the concept of establishing an event where all of the advancements of the preceding year would be recognized and celebrated with some form of achievement award has been a part of NWP's overall mission statement.

But in addition to the award show these are the other aspects of the "Shock The Monkey" celebration event:

A.) Shock Kids is an Alternative Energy Education Program for kids.

B.) Shock the Monkey Theme Park

This would mean that NWP acquire sufficient land for use as proving grounds for the new inventions. This acquired property will be referred to as, the land of "Oh-Zee."

And each year on this land there will be a global event called, "Shock the Monkey." The date of the event is not set, but could happen during the Christmas season.

C.) "Shock the Monkey" stuffed doll for fund-raising. Either for selling directly or to provide as a donation gratuity.

E.) Children's book: "Shock's Ring of Power," for fund-raising. Either for selling directly or to provide as a donation gratuity.

F.) "Shock the Monkey" Cartoon Series or Video Game.

One media attraction to this event will feature a huge TeePee that is decorated with multiple strings of colored Christmas lights that will be powered entirely by the ROP.

Additionally, regardless of if it happens concurrently with the "Shock the Monkey" annual event.

A major fund-raising event presently referred as, "The Truth Teepee," is being formulated. The goal here would be to build the largest ever teepee and illuminate it at night so it can be see from space at night. Also small teepee and other memorabilia can be sold to raise money for the decedents of the victims of the "**WOUNDED KNEE MASSACRE**" in December of 1890. The Peace-TeePee foundation will be formed for that purpose.

The "Shock the Monkey" event will also feature competitive games to produce the most electricity within time constraints or limited to number of participants. A highlight of the event will be an electric car race called

"Sparks 500."

The event concludes with an awards and recognition ceremony which will take place and be available to the general public.

Also during the month long event there will be seminars and topical talks presented. An example of what could be featured is the following example:

Earth is The Holy Land!

This one planet is the most sacred planet in the known universe. Mars may have had an atmosphere at one time or another. But Mars is no Earth!!! And to pretend as if moving from this planet to that totally inhospitable iron rock is just an inconvenience of distance is pure folly as well as sacrilegious. And then to alter or damage any of Earth's environment in order to do so is simply a crime against humanity if not against the earth.

But also in addition to improving man's lot on Earth. NWP will strive toward enhancing life for all of Earth's creatures. This is because all of NWP's activity will not pollute or destroy the habitats of the pristine wilderness.

When one considers that its been estimated that the Deepwater Horizon oil spill will ultimately cost BP upwards of $61 billion some skepticism should effect most to wonder how truly beneficial has our fossil fuel revolution really been. However even more detrimental, most spills are smaller with less impact and media scrutiny, but taken as aggregate of the total spills and leaks, we are truly estimating significant long-term damage to Earth's fragile eco-systems.

Clean-up and litigation are just a portion of the financial ramifications of an oil spill. So what Do the Cost of Oil Spills Include? Well here's a list, but are not limited to:

A.) Cleanup

B.) Containment

C.) Natural resource damage assessment and restoration

D.) Property Damage

E.) Litigation

F.) Mitigation

G.) Fines and penalties

H.) Public Relations

Who Pays For Oil Spills?

Well supposedly, the cost of an oil spill will not be paid by the public and will only be shouldered by those culpable. The Oil Pollution Act of 1990 essentially states that the party responsible for the oil spill must pay for cleanup operations and restoring natural resources.

The U.S. Coast Guard's Oil Spill Liability Trust Fund provides emergency funding for those involved in oil spill response and cleanup. The fund also covers the costs for natural resource damage assessment and restoration. When a party is deemed liable for the oil spill, they will be required to reimburse expenses paid by the fund.

But unfortunately there are always hidden costs to the

marine life destroyed or permanently damaged by any spill that simply removing the oil cannot resolve. So in the end we will still be left with that one gnawing question. Was it really worth it after all?

Chapter Five

A Launch For The Ages.

NWP is not about creating a new revenue stream fo a few billionaires while the rest, the multitude are left suckle off the trickle down. NWP is about everyone equally enjoying the wealth and prosperity that is generated by their collaborative efforts. And producing beneficial change in the societal malfeasance affecting us today.

So, that's why after careful consideration, NWP will be established as a Non-profit.

And this is the mission statement:

"To Cherish & Safeguard the Living Ecosystems that Adorn Our Most Sacred Earth by the Forces of Education, Wisdom & a Genuine Love of Life."

How will NWP operate as a non-profit by instituting the following:

1.) Shock the Monkey:

Plus in conjunction with the Shock the Monkey annual extravaganza event and awards show, there will be multiple on-going fund raising events:

A.) Website: The current website is intended to be rudimentary. However with the establishment of the non-profit, all of the pertinent and useful social media platforms will be used.

B.) Fund-raising: Currently all fund raising activity will revolve around the "Shock the Monkey"theme.

C.) Kids education: This will be focused on helping young children become aware of the alternators and generators ability to produce electricity. And how to work together and share the energy they produce.

D.) Land is to acquired for the "Proving Grounds" and where to establish home base for the "Shock the Monkey" annual event.

2.) Assist in the development of innovative electrical generating devices.

3.) Fund Promising Patents.

4.) License patents.

5.) Grants:

Once NWP is certified as a non-profit it will seek those grants deemed appropriate for its cause.

But in addition to grants and other fund-raising strategies, NWP will seek to highlight the enormous subsides now being doled out to corporations and multi-conglomerates that contribute the most to fossil fuel pollution.

Every year the oil industry is doled out billions in tax-payer subsides. While at the same time generating fantastic amounts of profit and pollution. This should be reprehensible, but in America and across the globe it's business as usual. NWP proposes to rectify that by utilizing those subsidies to counteract Big Oil's long-term destructive tendencies.

WHAT ARE THE FACTS?

Well, beginning by World War I, over one century ago, the U.S. government acted to stimulated oil and gas production in order to ensure a domestic supply.

And by 1995, Congress established the Deep Water Royalty Relief Act.10. Oil companies could now drill on federal property without cost of royalties.

And since 2011 these U.S. government oil industry subsidies have enabled a "gold rush" pursuit of Earth's precious ancient fluids.

1.) Volumetric Ethanol Excise Tax Credit - $31 billion.

2.) Intangible Drilling Costs - $8.9 billion.

3.) Oil and Gas Royalty Relief - $6.9 billion.

4.) Percentage Depletion Allowance - $4.327 billion.

5.) Refinery Equipment Deductions - $2.3 billion.

6.) Geological and Geophysical Costs Tax Credit - $698 million.

7.) Natural Gas Distribution Lines - $500 million.

8.) Ultradeepwater & Unconventional Natural Gas & other Petroleum Resources R&D - $230 mil.

9.) Passive Loss Exemption - $105 million.

10.) Unconventional Fossil Technology Program - $100 million.

11.) Other subsidies - $161 million.

A 2020 report by the "International Renewable Energy Agency" tracked some $634 billion in energy-sector subsidies in 2020, and found that around 70% went to fossil fuels.

Only 20% went to renewable power generation, 6% to biofuels and just over 3% to nuclear."

Finally, considering that the NWP vision quest attempts to remedy some portion of the polluting hazards caused by the fossil fuel industry, it would appear justifiable if however substantial a portion of those subsidies were allocated toward NWP's vision quest efforts.

Appendices

Appendix 1

This white paper was intended to provide additional mathematical proof that the theory of harnessing "Leveraged Momentum," is sound and practical.

Spiralogic: Sparking The Cycle Of Life Toward A New Age Of Energy Enlightenment.

©2018 Nicholas G Stangarone

This technical white paper provides a mathematical overview of an invention that has the potential to generate copious currents of clean moral electricity for the greater global good.

The equations presented will quantify the first manifestation of the electrical generating invention codenamed "Ring of Power." Basically, the device works as a force multiplier. That is, it transfers the leveraged momentum of speeding bicyclists into rotational motion. This harnessed force then acts to torque a speed of sufficient angular velocity to drive a cluster or bank of alternators.

The following four force-related formulas offer a quantitative appreciation of the power conversion capabilities of a leveraged momentum machine like the Ring of Power.

1.) Kinetic Energy

2.) Angular Momentum

3.) Torque and Leverage

Appendix 1

4.) Bike Gears and Pulley Transfer Ratios

However, we'll conclude by briefly touching on the socioeconomic ramifications of this invention.

1.) Kinetic Energy

Kinetic energy or the lack thereof is the reason why using a stationary bike to generate electric power is so woefully inefficient. Conversely, riding a bicycle out on the open road can seem so magically fantastic simply because it moves you through time and space. Today's bikes are basically highly efficient machines for transporting the human body over great distances at speeds generally faster than either walking or running.

The force or energy that is generated by the action of consistently rapid spatial displacements can be mathematically expressed as kinetic energy. That is, the equation is as follows:

$$KE = 1/2 \, I \, \omega^2$$

Where (I) represents the moment of Inertia for the object in motion.

(ω) is for angular velocity that the object is travelling.

So to delve further into why using a stationary bike is so inefficient for electrical production, let us compare the kinetic energy between spinning a bicycle wheel and a cyclist riding their bike on a level surface.

Appendix 1

To measure how much kinetic energy can be produced when a bike wheel is forced to rotate on its axle by pedalling, we'll begin with the dimensions of a standard 26-inch bicycle wheel rim that has a diameter of 559 mm (22.0") and an outside tire diameter of about 26.2" (665 mm). This correlates to an approximate wheel radius (r) that equals:

r = ((665 mm) + (559 mm) / 2) /2 = 306 mm or 0.306 m

Next, we'll complete the formula for measuring kinetic energy.

Wheel Weight = 2.3 kg

Inertial Constant (k) = 1

Moment of Inertia of Wheel:

I = (1) (2.3 kg) (0.306m)^2 = 0.22 kg m^2

The speed of the bicycle is 25 km/h (6.94 m/s; 15.5 mph).

The wheel's circular velocity is 3.32 revolutions/s

The angular velocity (ω) = 20.9 rad/s

The kinetic energy (E) of the spinning bicycle wheel is expressed in joules.

E = 0.5 (0.22 kg m-squared) (20.9 x 20.9 rad/s)

Therefore the kinetic energy of a spinning bike wheel equals approximately 47.9 J.

That is 47.9 joules.*

Appendix 1

Now let's quickly calculate the kinetic energy of a bicyclist travelling at 6.94 m/s (15.5 mph). The total mass of the lightweight cyclist and bike is 74 kg. (74 kg = 163 lbs)

Therefore, the kinetic energy of a travelling cyclist equates to 0.5 x 74 kg x 48 m/s = 1782 J.

That is 1782 Joules**

So the difference in kinetic energy between a cyclist riding their bike on a level roadway or the same person using a stationary bike is <u>1782 J - 47.9 J = 1,734 Joules.</u>

This is an additional equivalent force of 36 spinning bike wheels. In other words, this represents an additional 36 alternators to potentially produce that much more electrical power by the same cyclist.

Ideally, that individual cyclist would only exert identical amounts of physical intensity for each trial run. Also, the effects of thermal build-up from the increased caloric burn or the wind resistance experienced while travelling 15 miles per hour would be omitted from the overall energy matrix and as such, the preceding calculations clearly demonstrate why using a stationary bike to generate electricity is an extremely inefficient energy stratagem.

Again, the basic function of a bicycle is not the ability to rotate a bicycle wheel exceedingly fast. It is rather the practical utility of causing a human body to quickly traverse distances or in other words, rapid spatial displacement. That is the true power provided by this time-tested travel machine. And that is the source of power (kinetic energy)

Appendix 1

which must be harnessed to generate endless electrical energy.

2.) Angular Momentum

Before we calculate the angular momentum of the aforementioned cyclist as it pertains to the invention let's consider that the kinetic energy calculated for the travelling cyclist was measured as spatial displacement parallel to a straight line.

So using the previous quantities we can calculate the momentum as; p = mass x velocity.

74 kg. (74 kg = 163 lbs) x 6.94 m/s (15.5 mph) = 513.56 kg m/s***

And as previously indicated the rapid spatial displacement required to operate the "Ring of Power," is exclusively rotational. With a defined axis now available, the forces of torque and angular momentum can then be determined with precise accuracy.

Now, the angular momentum of a rigid object is defined as the product of its "moment of inertia" and its angular velocity. And as importantly, subject to the universal laws of physics and more specifically the conservation of angular momentum. ***

Appendix 1

Angular momentum L = I (Moment of Inertia) x ω (Angular velocity)***

I = MR² = 74 kg x (7.76 x 7.76) m = 4456 kg m²

L = 4456 kg m*m x 0.894 rad/s = 3984 kg m²/s

L = pD

L = 513.56 kg m/s x 7.76 m = 3985 kg m²/s

The dominant design characteristic of the "Ring of Power" is its circular configuration and consequently axis or pivot point to effect 360 degree rotational movement. This also means that located on its perimeter circumference there exists two points or contacts that are exactly 180 degrees apart. This will effectively function as provisional levers and will operate most efficiently when a pair of cyclists are positioned exactly at those two equally separated circumference contact points and effectively facing in opposite directions (left facing/right facing) and rather importantly, while travelling at or near identical speeds.

The duelling cyclists will then be positioned to equally contribute to the total force that will drive the alternators with nearly 8,000 kilograms of angular momentum.

But as will be shown in the next section, this force is used to torque double that amount or nearly 16,000 kilograms of real driving force!

Appendix 1

3.) Torque and Leverage

So now that we've demonstrated how the formidable kinetic energy of a speeding cyclist can result in a powerful primary mover for the Ring of Power. What we further hope to prove is that the true effectiveness of the invention is found in its ability to leverage the angular momentum of the dual cyclists into an amplified or multiplied torque.

First let's look at the equation for torque:

(Torque) τ = I (Moment of Inertia) multiplied by α (Angular Acceleration)

a = v^2 / r:

6.94 m/s (15.5 mph) and 7.76 m

$(6.94 \text{ m/s})^2$ / 7.76 m = 6.2 m/s^2

α = Angular Acceleration = a / r:

α = 6.2 m/s^2 / 7.76 m = 0.7998 rad/s^2

Torque = 4456 kg m^2 * 0.7998 = 3564 N m

As is indicated by the following diagram, with the duel cyclists positioned exactly 180 degrees apart, they will establish an imaginary pivot at point "C."

However, the torque that would have been experienced at point "C" is not imaginary but quite real and more importantly this force is conserved.

Appendix 1

Therefore this torque must be realised as an inverse-transfer to the opposing cyclist. Each cyclist produces a torque that the other cyclist experiences as a reduction in the resistance of the eight (min. configuration) alternators.

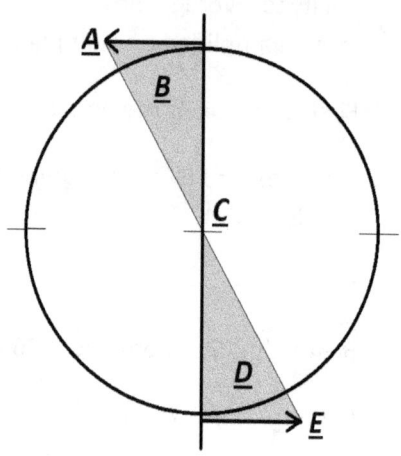

A.) Cyclist A travelling @ 15.5 mph (24.94 kph)

B.) Triangular Area of Inverse Torque (Cyclist B)

C.) Imaginary Pivot Point

D.) Triangular Area of Inverse Torque (Cyclist A)

E.) Cyclist B travelling @ 15.5 mph (24.94 kph)

4.) Pulley Transfer Ratios

Appendix 1

As previously mentioned, with duel cyclists pedalling their bicycles at an average speed of 15 mph, they should be able to generate enough power to drive a minimum of eight alternators with the invention's "Galaxy Cluster" formation.

Now, without having conducted any prior proto-type trial runs to verify certain assumptions we will currently rely on the following dimensions as structurally ideal until proven otherwise.

ROP Circumference = 160 feet (48.768 m)

ROP Radius = 25.46 feet (7.760208 m)

Transfer Pulley Circumference = 75 inches (1.905 m)

Transfer Pulley Radius = 12 inches (0.3048 m or 30.48 cm)

Alternator Pulley Circumference = 6.28 inches (0.159512 m or 15.9512 cm)

Alternator Pulley Radius = 1 inch (2.54 cm)

15 mph = 79,200 ft/hour = 1,320 ft/minute = 15,840 inch/minute

ROP = 8.25 rpm

Transfer Pulley = 211 rpm

Alternator pulley = 2,519.9 rpm (Note: 15,840 divided by 6.28 equals 2,522.2929)

Appendix 1

New World Power and the Promise of Good Energy

Throughout the ages, the greater mass of humanity has always lacked the strategic energies to truly effect far-reaching societal change, in spite of its greater aggregate population size.

Obviously the perturbed masses can be used to effect short term change by combining their overwhelming physical presence as in a mass demonstration or protest gathering.

That's because the total cumulative crowd is the real power which also immediately dissipates upon dispersal of the gathered individuals. Of course, the residual or echo responses created by the gathering can develop into their own movements. However these ancillary actions could also form counter to the original impetus.

It's reasonably understood that social media platforms offers fast and efficient connectivity between enormous numbers of "active users." But though these internet platforms are ultra-fast, it is also just part of a globally fermented stream of social media noise and useless distractions that easily inundate any single movement or cause in a cesspool of digital malaise. Basically, the social media platforms must be the instrument of the movement and not the movement, in and of itself.

And of course, there could always be an armed or quasi-military uprising and potentially an insurgent coup. Even possibly civil wars, but as has been shown by history, this course of action plays entirely into the authorities hands because that's their calculated expectation!

Appendix 1

Therefore, those in charge, unless totally incompetent, fully corrupt and or both are always expertly equipped and prepared to suppress any immediate threat of armed violence.

Let's be honest with ourselves, as long as the masses are hopelessly reliant on the wealth and power of a small minority they will always be like puppets on a string. In other words though you may feel like you're moving with deliberate initiative, in fact your actions are being guided by the stealthy strings of the ubiquitous puppet master.

And until the masses are able to cut the power cords of the global profit mongers by using their own wealth and power, the status quo of the money matrix will remain unquestioned.

However, New World Power respectfully begs to differ with this greed centric global-logic with a moral centric "spiral-logic." And it begins with the understanding that the masses should never expect the "1%" to act in any fiduciary capacity toward the benefit of mass humanity. For they will surely be greatly disappointed.

Big money will always act in whatever manner optimizes their return on investment or profit. They will always act for the further benefit of maintaining, if not enriching their wealth, fortune and status quo.

Plain and simple, their profit is their prophet.

NWP believes that it's solely incumbent upon the indentured masses to create their own power and profit and without

Appendix 1

strings attached. But more importantly, it would behoove the masses to act on a strategic and coordinated causeway without resorting to violence.

For in the end, should a military insurgency succeed in removing the greedy few, they will be constrained by their own violent actions to choose previously like-minded personnel to replace those they forcibly removed. Consequently the "Energy Enlightenment" necessary to effect a true social transformation to a higher moralistic state of consciousness will prove unattainable.

Growing up in post-ww2 America, we were throughly indoctrinated to believe that we lived in the most perfect society ever imagine by man. And it would not be long before that modernity would deliver us to that promised land of the future. A time and space where we would be able to fly our cars anywhere we wished and all tedious work would be delegated to totally obedient robots.

While in our digitized virtual worlds we may have fooled ourselves into believing we have reached our promised land. Even so, in our unalterable determination to live this life of fantasy that's so unrealistically far beyond our means, we may have, quite unfortunately, established an enduring false sense of entitled prosperity. And in our blind faith to maintain this mass state of mind-numbing delusion we have reduced ourselves to stealing the resources of future generations with a despicably callous abandon.

Appendix 1

Arguably, we have stolen from the generations destined to come, these absolutely helpless generations of tomorrow, their planet Earth. Once, a truly living paradise; alone in a vast lifeless universe. Unbelievably abundant of amazing wildlife and life sustaining vegetation that only cherished praise on their glorious life-giving creator. Now we will leave them unsustainable skyscrapers, a continent worth of asphalt parking lots linked together by endless miles of cement highways. Not to mention the billions of four wheeled gas guzzlers racing over them every single day.

And so, we have with a greedy malice stolen from coming generations their rightful claim or fair share as future inhabitants of this planet, or more specifically, from Earth's profoundly unique legacy; her ancient oceans petroleum or hydro-carbons molecules.

The NWP is a promise to provide the moral and electrical power necessary to successfully reevaluate this greedy construct before it is far too gone.

Also the "Ring of Power" technology only exists to rescue mankind from the greedy tyranny of the petrodollars power matrix; thereby giving man a new means of righteous energy endeavour.

But unfortunately, there really isn't much time, since within short order, robots not only be designed and built to replace most employees at their current jobs will flood the marketplace. But more ominously, robots also hold the technological prowess to become the most awesome and effective fighting machine in the history of human warfare.

Appendix 1

Sooner or later, warbots will be coming to a battle front near you.

Will you be ready? Just kidding of course, because once that day arrives it will already be too late.

* Calculations provided by engineeringtoolbox.com (11/6/2018)

** Calculations provided by answers.yahoo.com (11/14/2018)

***http://hyperphysics.phy-astr.gsu.edu/

Appendix 2

Original Patent Idea

Several years ago I completed a provisional patent application for the original model. Still circumstances would not allow to proceed as planned and it was put on the back burner once more.

But in honor of the late great Steve Jobs, I purposely waited until the month of October, or more specifically, Wednesday, October 1, 2014, to submit a United States Provisional Patent application for the "Ring of Power." So I can therefore state that by October 5th, the invention was patent pending.

Though this invention was never intended to be a working model. The primarily objective here was to establish preeminence in being the first to suggest in practical terms the means to which Leverage Momentum could be harnessed. The important aspects of the provisional patent application are as follows:

Brief Summary of the Invention

The **Cyclonic Cylinder STS (Speed Transfer System)** is a leveraged momentum machine that transfers or converts the speed of the traveling cyclists into the sufficient rpm of the alternator rotor to produce electrical power.

Basically, a minimum of four to six PowerWheels are position at equal distance from a center point and also at equal spacing from each other. Each PowerWheel serves

Appendix 2

three purposes. First they are to act as supports for the integrity of the circular form. Secondly, there are to secure at least one alternator to the post and the power wheel pulley. Thirdly, they are to transfer the rotational speed of the Transfer Cylinder to the alternator rotor.

Once the power wheels are secured and the transfer cylinder assembled, the **Cyclonic Cylinder STS (Speed Transfer System)** can be operated by a minimum of two bicyclists riding in tandem and positioned at opposite ends of the Transfer Cylinder.

The momentum of the bicyclists is leveraged on the transfer cylinder by the rapid displacement of their polar positioning (180 degree separation) on the transfer cylinder. Also the flywheel effect of the Transfer Cylinder coupled with the angular momentum of the bicyclists will act to maintain the acquired speed.

Riding in tandem to maintain a speed between 15-30 mph, revolves the Transfer Cylinder at the optimal speed to spin the minimum number of four alternators at the rpms necessary to generate electricity of sufficient voltage and current to power the basic needs of its users.

In effect, by riding in tandem, a constant speed can be maintained for extended periods and more efficiently than riding individually.

Appendix 2

Brief Description of the Drawings

Drawing No.1: *Diagram of the Cyclonic Cylinder STS (Speed Transfer System).*

Figure 1 shows diagram of the **Cyclonic Cylinder STS (Speed Transfer System)** in the basic configuration. This basic configuration requires a minimum number of PowerWheels (4) and bicyclists (2).

   1.) Transfer Cylinder
      2.) PowerWheel with attached Alternator
   3.) Standard Bicycle

Drawing No.1

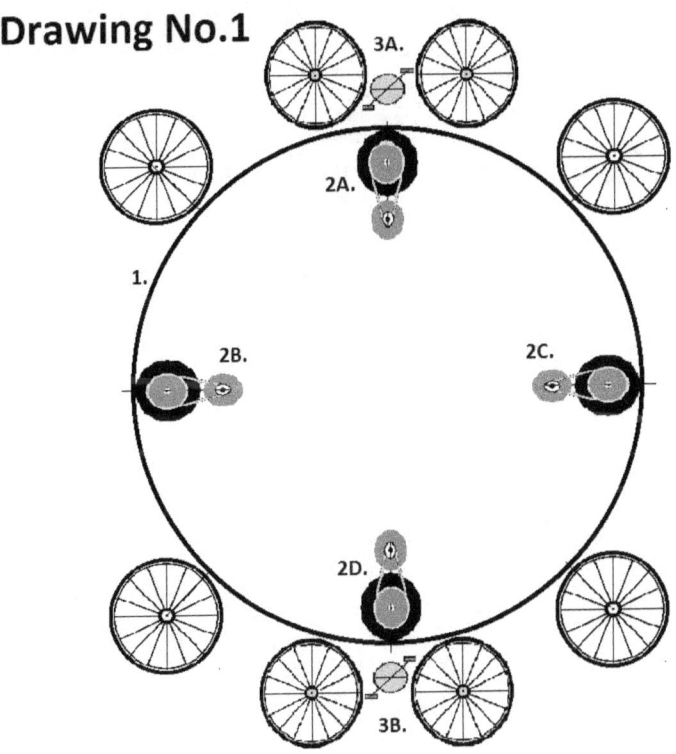

Appendix 2

Drawing No.2: *Power Wheel without alternator*

Shows a side view of the Power Wheel without the alternator.

A.) 1' dia. Pneumatic tire
B.) Axle and attaching cap
C.) 1' dia. Pulley
D.) Adjustable alternator harness
E.) 2 guide lines with brackets
F.) 3.5' fence pole
G.) 3.0' support post

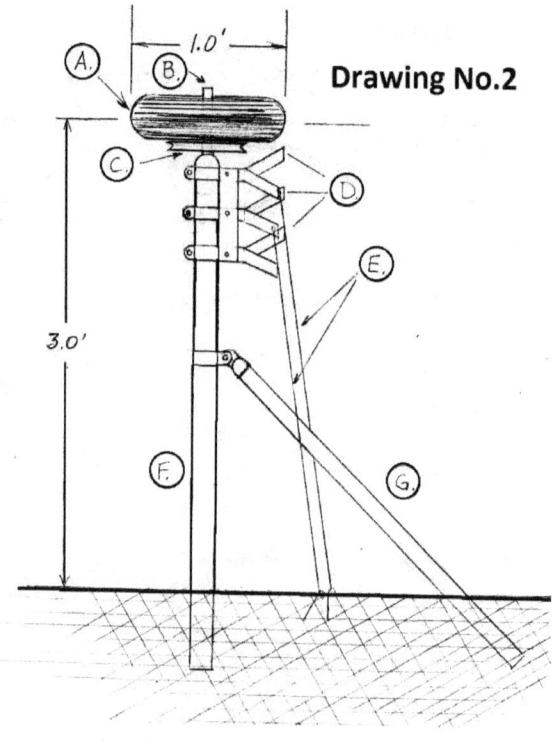

Drawing No.2

Appendix 2

Drawing No.3: *Wheelbend Quadtrain*

Figure 1 show a top view of the Wheelbend section.

Figure 2 shows a front view of the Wheelbend section.

Figure 3 shows a side view of the Wheelbend section.

Each Wheel Quadtrain consists of the following:

A.) Fabricated 10x1' sheet metal cylinder section

B.) 10' long, 2" diameter structural support steel tubing

C.) 26" bike wheel

D.) 26" front fork

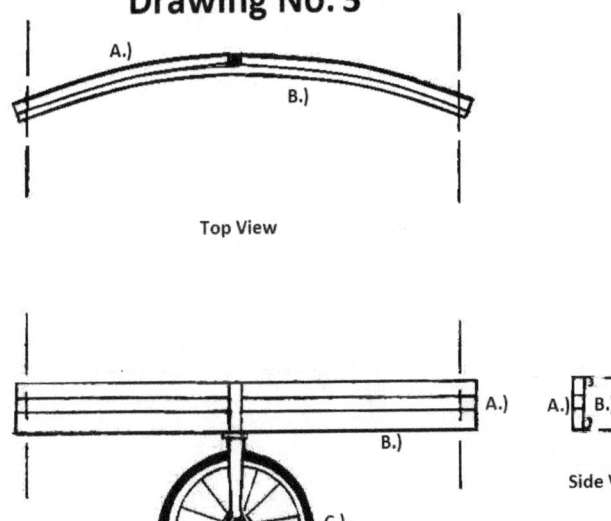

Appendix 2

Drawing No.4: _Hoopbend Quadtrain_

Figure 1 show a top view of the Hoopbend section.
Figure 2 shows a front view of the Hoopbend section.
Figure 3 shows a side view of the Hoopbend section
Each Hoopbend Quadtrain consists of the following:

A.) Fabricated 10x1' sheet metal cylinder section
B.) 10' long, 2" diameter structural support tubing
C.) 6" dia. Metal Hoop Ring

Drawing No. 4

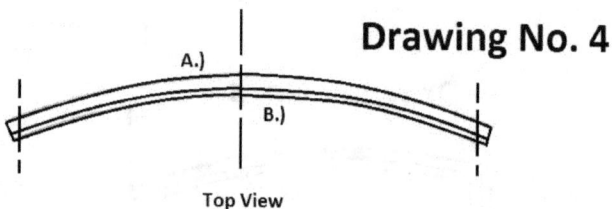

Top View

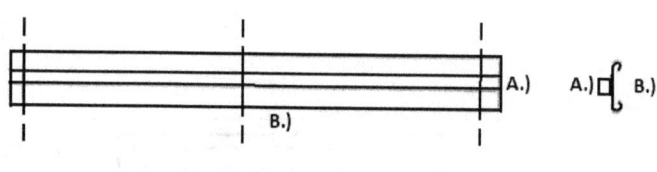

Front View

Appendix 2

Drawing No.5: *<u>Bike to Transfer Cylinder Connectors and Lock Bracket</u>*

Figure 1 show a side view of the Bike to Transfer Cylinder connector

Figure 2 shows a side view of the steel Hoop

Figure 3 shows a side view of the Bike Seat connector   **Figure 4** shows a side and front view of the Lock Bracket

A.) Large Clasp
B.) Small Clasp
C.) Rope
D.) Shows a side view of the steel Hoop
E.) Shows a side view of the Bike Seat connector
F.) Shows views of the 2x2x6" U-Lock Steel Bracket

Drawing No.5

A.)
B.)
C.)
D.)
E.)
F.)
Side View
Front View

Appendix 2

Detailed Description

(This basic configuration is the smallest version of the **Cyclonic Cylinder Speed Transfer System** design.)

Drawing No.1 shows a diagram of the Cyclonic Cylinder STS (Speed Transfer System) in its basic configuration. This basic configuration requires a minimum number of four PowerWheels and two bicyclists to operate efficiently.

The basic configuration of Transfer Cylinder consists of four Wheelbend sections (**Drawing No. 3**) and four Hoopbend sections (**Drawing No. 4**). Each of the eight sections measure ten feet in length. When fully assembled, the Transfer Cylinder measures eighty feet in circumference.

The device leverages the momentum of speeding bicyclist by attaching the bicycle by means of a specified length of rope or strapping to a rotating sheet metal cylinder. This cylinder rotates on a minimum of four bicycle wheels. The large diameter cylinder is held in place by a minimum of four power wheels.

Drawing No.2 shows the power wheels with reinforced posts serves three specific functions. First they are placed at equal distance from center to form a circular perimeter of specified circumference to sufficiently stabilize the integrity of a rapidly rotating sheet metal cylinder. Secondly, it must secure one alternator to itself for belt and pulley harnessing. Thirdly, the power wheel operates with the transfer cylinder to transfer the kinetic energy of the bicyclist's angular

Appendix 2

momentum and speed to the angular velocity of the alternator's rotor.

The elimination of spokes for stabilization of the rotating Transfer Cylinder reduces the area required for utilization. This design feature creates usable space that may be occupied by certain structures or activities. For example, the space could be occupied by a small building or storage facility. Other suitable structures could be the geodesic dome or teepee. Also a tower for electrical distribution could also be fitted within the circular area of the unused space.

As the bicyclists ride at the optimal speed, the rotating Transfer Cylinder transfers the speed of tandem bicyclists to the PowerWheels, which in turn spin alternators by a belt and pulley to generate electricity. Most importantly, the **Cyclonic Cylinder STS (Speed Transfer System)** uses power of the speeding bicyclist's leveraged momentum to maintain the optimal speed for the longest possible time. The leverage is realized by the motion of opposing bicyclists traveling in a clockwise or counter-clockwise direction to generate the torque necessary to rotate the Transfer-Cylinder

Dimensions of the
Cyclonic Cylinder STS (Speed Transfer System):

Circumference: 80 feet
Radius: 12.73 feet
Height: approx. 3.5 feet

Appendix 2

Assembly Instructions.

Step #1.) Determine suitable location to establish the point of center.

For the area to be suitable for the safe operation of the **Cyclonic Cylinder STS (Speed Transfer System)**, it must be free of obstacles like rocks and shrubs. The suitable area must also be relatively level and flat plane. Also for aesthetic consideration, the ideal area would be situated as

near to scenic surroundings as practical.

The transfer cylinder requires a 25' diameter and the bicyclists require additional footage or about four additional feet. That means, you'll need at least 29'-30' square foot area.

Step #2.) Once a location has been determined, find the center or pivot point.

From this point, measure out to locate the positions for each PowerWheel posts.

In addition to locating the equal point from center for each PowerWheel. The PowerWheels must also be distanced equally from each other.

Step #3.) Set the PowerWheel support post in designated location.

Each PowerWheel support post should be secured using the same methods and techniques to properly set steel fence posts. All play or wobble in the post setting

Appendix 2

should be minimized if not eliminated. If needed, additional guide wire can be used to tension up the setting.

Step #4.) Assemble PowerWheel.

Step #4A.) Attach alternator harness to post.

(See Drawing No.2: Power Wheel without alternator Figures A, B & C for the following.)

Step #4B.) Attach axle to post.

Step #4C.) Secure one foot dia. Rubber Tire with attached pulley.

(See Drawings 3 and 4 for the following.)

Step #4D.) Assemble Transfer-Cylinder. Place all eight fully assembled sections around the perimeter formed by the power wheel posts. Secure the sections with the metal brackets.

Step #4E.) Assemble alternators and connect to junction box. Secure one alternator to each post with harness. Attach wiring from alternator to junction box.

Step #4F.) Attach bicycles. Secure strap around seat post. Using the (x) foot length rope to tie the bike and then to tie the transfer cylinder rings.

Step #4G.) Begin cycling power. Monitor power for consumption and transmission. The best mode for the smallest version, which is 80' circumference, is between two and four bicyclists, riding in tandem, and operating eight power wheel posts. The bicyclist should maintain a speed between 15 mph to 30 mph.

This is only the first model, and as such subject to

Appendix 2

profound improvements and refinements. But this is more than just an engineering innovation, because if it proves as effective as theorized it could form the nucleus for a social movement that will free the indentured multitude from the rapacious greed of the energy hegemony.

Appendix 3

This is a reprint of Chapter 4 from the book titled, "The Ring of Power."

The New Ordered World is Now.

It's becoming increasing difficult to disguise the fact that a majority of this planet's human population now finds itself in a deplorable state of collective servitude. For everyday, this vast multitude is expected to pay out costly sums in fees and taxes! Even worse, this mass state of indentured exploitation by the insatiable greed of energy profiteers has festered and grown more onerous since it's dubious inception during the age of steam.

Even Nikola Tesla, the electrical wizard who single-handedly illuminated the 1893 Chicago Exhibition with his patented and highly efficient alternators, had railed against this greedy construct. And remained determined and steadfast in his personal quest toward discovering a free source of electrical power that would end the monopolistic reign of the utilities and their unabated exploitation of the uninformed masses to the end of his life.

But why take my word for it! Read for yourself the

Appendix 3

following passages from an article that appeared in the **March-1896** edition of the World Sunday Magazine that was aptly titled, "Earth Electricity To Kill Monopoly."

Some would say this harkens of a clarion call demanding some form of utility debt relief for the working masses.

For the opening paragraph unabashedly declares, "The end has come to telegraph and telephone monopolies with a crash...all the other monopolies that depend on power of any kind will come to a sudden stop."

It's almost comical at how absolutely dead wrong the author was proven to be, especially when considering our deplorable servile condition, over a century later.

But this was just a tame prelude to the very next paragraph that stated unequivocally, "It means that if Nikola Tesla succeeds in harnessing the electrical earth currents and putting them to work for man there will (be) an end to oppressive, extortionate monopolies in steam, telephone, telegraphs..and the grasping millionaires who have for two decades milked the people's purse with electrical fingers..."

A few paragraphs down, the article ends with a few more revealing tidbits, "Electricity will be as free as the air. For the privilege of its use legislatures will not have to be bribed or men corrupted at the polls, and public boards will not have to be seen to bestow exclusive franchises upon corporations organized to use public property for purposes of private gain, and make the people pay the original cost of their investment and

Appendix 3

excessive charges for service...Monopolies for purveying steam power too will be forced to capitulate to free electricity...The successful adaptation of Tesla's discovery will administer a death-blow to the most galling slavery that has yoked the activities of men to the treadmill of monopoly."

The Ring of Power draws from the sake of posterity, Tesla's gleaming alternating current aspirations to produce plentiful power for all humanity. Specifically, the Ring of Power is a "leveraged momentum machine" and combines most of the known forces in the Universe to deliver electrical power to those who can least afford it.

Let me reiterate for unmistakable clarification, human beings have been endowed with the capacity to be energy self-sufficient. It is only through the artificial constraints of a greedy society that we find ourselves forced to owe our existence to these global, energy-hoarding profiteers. We have been cajoled and berated into accepting the false premise that the ability of man to create their own power is a futile attempt at best and should therefore be avoided at all costs.

These money-minded disciples of greed are quick to throw out comparative energy costs and nascent technology limitations. They will throw calculations in your face to demonstrate just how relatively little a human can generate with current technologies. But this is all just what it is, words from the master declaring that everything is fine just as they are. And for the sake of

Appendix 3

the burgeoning global economy, we should keep our ignorant, cotton-picking hands out of their highly lucrative free markets.

And that's why it's incumbent upon those not fully-invested with this lopsided socioeconomic construct to make the necessary efforts to rediscover this great potential of the human body. And it's not simply about leveraging muscle power, it's about leveraging momentum. You know, the reason why collisions in full contact sports like football and hockey can be so devastating.

And it's with this thought in mind that I propose the following calculation: People-Power equals Clean Energy plus Clean Environment multiplied by a healthier and wealthier population as the sum numerator. Then divide this sum by a shrinking unemployment and national deficit.

What you would ultimately end up with should finally equate to a truly financially and physically sound nation. Of course, the goal is for the denominator to shrink to zero and then cancel itself out.

But in all seriousness, how can we make this calculation a reality and not just bold words inked on paper? To answer that I would offer the following observation. We are fast approaching the fiftieth anniversary of man's first visit to moon. But fifty years before that historic Apollo Moon Launch, the mere thought of going to the moon was considered fanciful science fiction. Now, it's just a matter of time before we

Appendix 3

visit our crimson celestial neighbor, the planet Mars.

But brain power does not preclude our daily requirements of physical activity. Its not by chance that our minds are enhanced and leverage by our capacity to manifest willpower through our primary force actuators. That is, our legs, arms, hands and feet.

Yet, once again, by circumstance of man's own contrived mechanical and electrical devices, social norms, taboos and other egregious forms of societal malfeasance the masses have been made to believe that we must pay to stay fit and healthy.

This cost to society is now accepted as a way of life. But what if we could turn the table on this greedy calculation and pay people for staying physically fit and healthy. Would they not be more productive and happier? And wouldn't the money they earned ultimately circulate back into economy?

But where would this money come from to pay people to stay fit, trim and healthy? Well, basically, the money comes from the power they generate. Which is why I truly believe that it's not an overstatement to claim that the New World Power movement has the potential to genuinely free the masses from the shackles of costly energy and establish a New Ordered World (NOW).

To do this, NWP will offer powerful incentives to motivate it's participants to become fit and <u>stay in shape!</u> This long term lifestyle change will be accomplished through generous monetary and recognition rewards for their exercise efforts. But this is only the start of what

Appendix 3

New World Power hopes to achieve for the betterment of our global society and not just a privileged few.

For one of the more ambitious goals of NWP will be to offer the able-bodied unemployed and or the homeless the opportunity to earn gainful wages through wholesome, body-building exercise. But again, the movement will not stop there because NWP, in conjunction with its "Fight Fat & Live Fit," program, will offer healthy and wholesome meals at little or no cost to all NWP members. Also in addition to this, NWP plans to provide onsite access to personal trainers, physical therapists and medical experts who can provide a full spectrum of healthy living expertise. And while it may not fall within the scope of NWP, we will work with local, state and federal agencies to see that these people will have full access to other housing, employment and educational opportunities.

Contained in the New World Power movement exists the momentous spark for the societal transformations capable of evolving our society beyond ways presently imaginable. This growing confidence stems from the knowledge that people should have the means to be energy self-supportive and truly free from the tyrannical greed of global corporate governance.

To achieve NOW, the New World Power movement, like any worthwhile effort that truly benefits mankind will be an epic struggle. But it is a good fight worth fighting because while however lofty may be NWP's goals and objectives, once reached, these transformations

Appendix 3

will unleash a renewed wave of hope in humanity as a noble collective soul imbued of divine aspirations.

So how does NWP plan to wean our society from it's destructive addiction to expensive foreign energy sources? First off, NWP will not rely on any one product, technoloy nor strategy because it will take everything necessary for the effective realization of NOW.

However at the time of this writing, the lead objective calls for obtaining adequate funding to construct a working prototype of the next generation, "Ring of Power." And also to conduct a series of tests and trials for determining the currently unknown, though theorized, operating parameters of this new device for generating electricity.

If everything moves forward accordingly, we should expect to see the first trials for the working prototype begin no later than July 16, 2015.

And by the 50th anniversary of Apollo Moon landing, the "Ring of Power," will become a stabilizing force in the global affairs of our planet Earth.

Appendix 4

Here is an example of a possible "Shock the Monkey" Event Extravaganza.

www.ingramcontent.com/pod-product-compliance
Lightning Source LLC
Chambersburg PA
CBHW050247220526
45465CB00002B/585